HVAC - Variable Refrigerant Flow (VF

HVAC - Variable Refrigerant Flow (VRF)

VRF (Variable refrigerant flow) is an air-condition system configuration where there is one outdoor condensing unit and multiple indoor units. The term variable refrigerant flow (VRF) refers to the ability of the system to control the amount of refrigerant flowing to the multiple evaporators (indoor units), enabling the use of many evaporators of differing capacities and configurations connected to single condensing unit. The arrangement provides an individualized comfort control, and simultaneous heating and cooling in different zones.

Currently widely applied in large buildings especially in Japan and Europe, these systems are just starting to be introduced in the U.S. The VRF technology/system was developed and designed by Daikin Industries, Japan who named and protected the term variable refrigerant volume (VRV) system so other manufacturers use the term VRF "variable refrigerant flow". In essence both are same.

With a higher efficiency and increased controllability, the VRF system can help achieve a sustainable design. Unfortunately, the design of VRF systems is more complicated and requires additional work compared to designing a conventional direct expansion (DX) system.

This 3-hour quick book provides an overview of VRF system technology. Emphasis is placed on the control principles, terminology, basic components, advantages and design limitations. This course is aimed at the personnel who have some limited background in the air conditioning field and is suitable for mechanical, electrical, controls and HVAC engineers, architects, building designers, contractors, estimators, energy auditors and facility managers.

HVAC - Variable Refrigerant Flow (VRF) Systems

The course includes a multiple-choice quiz consisting of fifteen (15) questions at the end.

Learning Objective

At the conclusion of this course, the reader will:

- Understand the difference between multi-split air conditioning system and VRF systems;
- Understand the operating principle of direct expansion split and VRF system;
- Understand the concept of thermal zone;
- Understand how VRF with heat recovery are different from ordinary heat pump systems;
- Understand the operation of thermostatic expansion valve (TXV) and electronic expansion valve (EEV);
- Understand the influence of building characteristics and load profile on selection of VRF system;
- Learn the advantages and application of VRF systems;
- Understand the design limitations and challenges in design of VRF systems.

HVAC - Variable Refrigerant Flow (VRF) Systems

OVERVIEW OF VRF SYSTEMS

The primary function of all air-conditioning systems is to provide thermal comfort for building occupants. There are a wide range of air conditioning systems available, staring from the basic window-fitted unit to the small split systems, medium scale package units, large chilled water systems and very latest variable refrigerant flow (VRF) system.

The term VRF refers to the ability of the system to control the amount of refrigerant flowing to each of the evaporators, enabling the use of many evaporators of differing capacities and configurations, individualized comfort control, simultaneous heating and cooling in different zones, and heat recovery from one zone to another. VRF systems operate on the direct expansion (DX) principle meaning that heat is transferred to or from the space directly by circulating refrigerant to evaporators located near or within the conditioned space. Refrigerant flow control is the key to many advantages as well as the major technical challenge of VRF systems.

Note the term VRF systems should not be confused with the centralized VAV (variable air volume) systems, which work by varying the air flow to the conditioned space on variation in room loads.

Split Air-conditioning Systems

Split type air conditioning systems are one to one system consisting of one evaporator (fan coil) unit connected to an external condensing unit. Both the indoor and outdoor unit are connected through copper tubing and electrical cabling.

The indoor part (evaporator) pulls heat out from the surrounding air while the outdoor condensing unit transfers the heat into the environment.

HVAC - Variable Refrigerant Flow (VRF) Systems

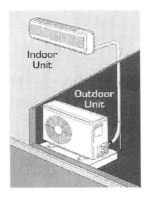

Split Air-conditioning System

Advantages of using Split Air conditioners

- Low initial cost, less noise and ease of installation;
- Good alternative to ducted systems;
- Each system is totally independent and has its own control.

Disadvantages

- There is limitation on the distance between the indoor and outdoor unit i.e. refrigerant piping can't exceed the limits stipulated by the manufacturer (usually 100 to 150 ft) otherwise the performance will suffer;
- Maintenance (cleaning/change of filters) is within the occupied space;
- Limited air throw, which can lead to possible hot/cold spots;
- Impact on building aesthetics of large building because too many outdoor units will spoil the appearance of the building.

Multi Split Systems

A multi type air conditioning system operates on the same principles as a split type air conditioning system however in this case there are 'multiple' evaporator units connected to one external condensing unit. These simple systems were designed mainly for small to medium commercial applications where the installation of ductwork was either too expensive, or aesthetically unacceptable. The small-bore refrigerant piping, which connects the indoor and outdoor units requires much lower space and is easier to install than the metal ducting. Each indoor unit has its own set of refrigerant pipe work connecting it to the outdoor unit.

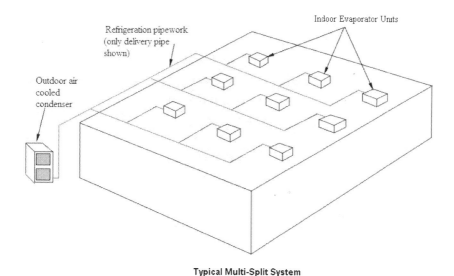

Typical Multi-Split System

Advantages of Multi-splits

- The fact that one large condenser can be connected to multiple evaporators within the building reduces and/or eliminate the need for ductwork installation completely.

- Multi-splits are suitable for single thermal zone* applications with very similar heat gains / losses.

Drawbacks

HVAC - Variable Refrigerant Flow (VRF) Systems

- Inability to provide individual control;
- Multi-split systems turn OFF or ON completely in response to a single thermostat/control station, which operates the whole system. These systems are therefore not suitable for areas/rooms with variable heat gain/loss characteristics.

*Thermal zone: A thermal zone is referred to a space or group of spaces within a building with similar heating and cooling requirements. Each thermal zone must be 'separately controlled' if conditions conducive to comfort are to be provided by an HVAC system.

Any area that requires different temperature, humidity and filtration needs shall be categorized as an independent zone and shall be controlled by dedicated control or HVAC system. Few examples below illustrate and clarify the zone concept:

- A conference room designed for 50 people occupancy shall experience lower temperatures when it is half or quarterly occupied. The design thus shall keep provision for a dedicated temperature controller for this zone;
- A smoking lounge of airport has different filtration, ventilation (air changes) and pressure requirement compared to other areas therefore is a separate zone;
- A hotel lobby area is different from the guest rooms or the restaurant area because of occupancy variations;
- In a commercial building, the space containing data processing equipment such as servers, photocopiers, fax machines and printers see much larger heat load than the other areas and hence is a different thermal zone;

- A hospital testing laboratory, isolation rooms and operation theatre demand different indoor conditions/pressure relationships than the rest of areas and thus shall be treated as a separate zones;
- A control room or processing facilities in industrial set up may require a high degree of cleanliness/positive pressure to prevent ingress of dust/hazardous elements and thus may be treated as separate zone.

Variable Refrigerant Flow or VRF Systems

VRF systems are similar to the multi-split systems, which connect one outdoor section to several evaporators. However, multi-split systems turn OFF or ON completely in response to one master controller, whereas VRF systems continually adjust the flow of refrigerant to each indoor evaporator. The control is achieved by continually varying the flow of refrigerant through a pulse modulating valve (PMV) whose opening is determined by the microprocessor receiving information from the thermistor sensors in each indoor unit. The indoor units are linked by a control wire to the outdoor unit, which responds to the demand from the indoor units, by varying its compressor speed to match the total cooling and/or heating requirements.

VRF systems promise a more energy-efficient strategy (estimates range from 11% to 17% less energy compared to conventional units) at a somewhat higher cost.

HVAC - Variable Refrigerant Flow (VRF) Systems

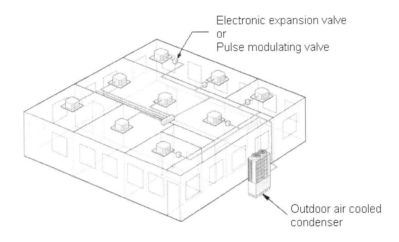

VRF System with Multiple Indoor Evaporator Units

The modern VRF technology uses an inverter-driven scroll compressor and permits as many as 48 or more indoor units to operate off one outdoor unit (varies from manufacturer to manufacturer). The inverter scroll compressors are capable of changing the speed to follow the variations in total cooling/heating load as determined by the suction gas pressure measured on the condensing unit. The capacity control range can be as low as 6% to 100%.

Refrigerant piping runs of more than 200 ft are possible, and outdoor units are available in sizes up to 240,000 Btuh.

A schematic VRF arrangement is indicated below:

HVAC - Variable Refrigerant Flow (VRF) Systems

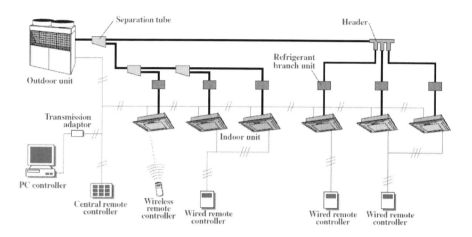

Figure (Source: Fujitsu)

VRF systems are engineered systems and use complex refrigerant and oil control circuitry. The refrigerant pipe-work uses number of separation tubes and/or headers (refer schematic figure above).

Separation tube has 2 branches whereas header has more than 2 branches. Either or both the separation tube and header can be used for branches, but the separation tube is **NEVER** provided after the header because of balancing issues.

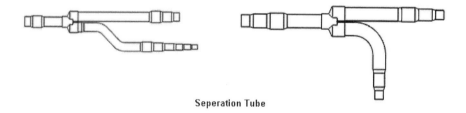

Seperation Tube

HVAC - Variable Refrigerant Flow (VRF) Systems

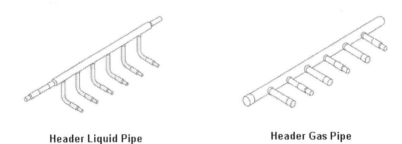

Header Liquid Pipe   Header Gas Pipe

Compared to multi-split systems, VRF systems minimize the refrigerant path and use **less** copper tubing. Minimizing refrigerant path allows for maximizing efficiency of refrigerant work.

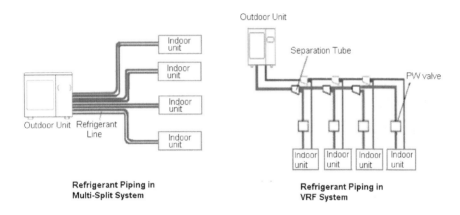

Refrigerant Piping in Multi-Split System   Refrigerant Piping in VRF System

Types of VRF

VRV/VRF systems can be used for cooling only, heat pumping and heat recovery. On heat pump models there are two basic types of VRF system: heat pump systems and energy-recovery.

<u>VRF heat pump systems</u>

VRF heat pump systems permit heating in all of the indoor units, or cooling in all the units, but **NOT** simultaneous heating and cooling. When the indoor

10

units are in the cooling mode, they act as evaporators; when they are in the heating mode, they act as condensers. These are also termed two-pipe system.

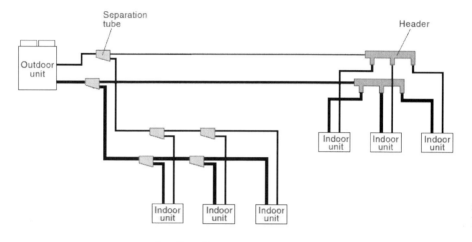

Cooling Type VRF System

Figure (Source: Fujitsu)

VRF heat pump systems are effectively applied in open plan areas, retail stores, cellular offices and any other areas that require cooling or heating during the same operational periods.

Heat Recovery VRF system (VRF-HR)

Variable refrigerant flow systems with heat recovery capability allow indoor units to operate simultaneously in heating and/or cooling mode, enabling heat to be used, rather than rejected as it would be in traditional heat pump systems.

VRF-HR systems also build on traditional heat pump technology by integrating inverters to drive compressors and fans and sophisticated expansion valve, refrigerant and distributed control are an enhanced version of the VRF heat pump and operate in net heating or net cooling mode depending on which is in greater demand by the conditioned space.

Each manufacturer has its own proprietary design (2-pipe or 3-pipe system), but most uses a three-pipe system (liquid line, a hot gas line and a suction line) and special valving arrangements. Each indoor unit is branched off from the 3 pipes using solenoid valves. An indoor unit requiring cooling will open its liquid line and suction line valves and act as an evaporator. An indoor unit requiring heating will open its hot gas and liquid line valves and will act as a condenser.

Typically, extra heat exchangers in distribution boxes are used to transfer some reject heat from the superheated refrigerant exiting the zone being cooled to the refrigerant that is going to the zone to be heated. This balancing act has the potential to produce significant energy savings.

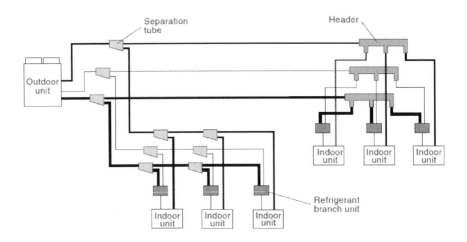

Heat Recovery Type VRF System

Figure (Source: Fujitsu)

VRF-HR mixed mode operation leads to energy savings as both ends of the thermodynamic cycle are delivering useful heat exchange. If a system has a cooling COP (Coefficient of Performance) of 3, and a heating COP of 4, then heat recovery operation could yield a COP as high as 7. It should be noted that this perfect balance of heating and cooling demand is unlikely to occur for many hours each year, but whenever mixed mode is used energy is saved. In mixed mode the energy consumption is dictated by the larger demand, heating or cooling, and the lesser demand, cooling or heating is delivered free. Units are now available to deliver the heat removed from space cooling into hot water for space heating, domestic hot water or leisure applications, so that mixed mode is utilized for more of the year.

VRF-HR systems work best when there is a need for some of the spaces to be cooled and some of them to be heated during the same period; this often occurs in the winter in medium-sized to large sized buildings with a substantial core or in the areas on the north and south sides of a building.

*COP – Performance rating used primarily in heat pumps. The Coefficient of Performance - COP – is defined as the ratio of heat output to the amount of energy input of a heat pump. It compares the heat produced by the heat pump to the heat you would get from resistance heat. COPs vary with the outside temperature: as the temperature falls, the COP falls also, since the heat pump is less efficient at lower temperatures. ARI standards compare equipment at two temperatures, 47°F and 17°F, to give you an idea of the COP in both mild and colder temperatures.

Refrigerant Modulation in VRF System

VRV/VRF technology is based on the simple vapor compression cycle same as conventional split air conditioning systems, but give you the ability to continuously control and adjust the flow of refrigerant to different internal units, depending on the heating and cooling needs of each area of the

HVAC - Variable Refrigerant Flow (VRF) Systems

building. The refrigerant flow to each evaporator is adjusted precisely through pulse wave electronic expansion valve in conjunction with inverter and multiple compressors of varying capacity in response to changes in the cooling or heating requirement within the air conditioned space.

We will discuss this further but before that let's refresh basic refrigeration cycle.

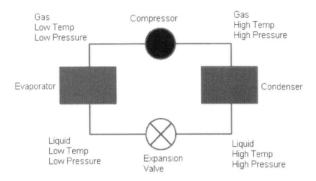

The fundamental of an air conditioning system is the use of a refrigerant to absorb heat from the indoor environment and transfer it to the external environment. In the cooling mode, indoor units are supplied with liquid refrigerant. The amount of refrigerant flowing through the unit is controlled via an expansion valve located inside the unit. When the refrigerant enters the coil, it undergoes a phase change (evaporation) that extracts heat from the space, thereby cooling the room. The heat extracted from the space is exhausted to ambient air.

Refrigeration systems can operate on <u>reverse cycle</u> with an inclusion of special 4-way reversing valve, enabling the absorption of heat from the external environment and using this heat to raise the internal temperature. When in the heating mode, indoor units are supplied with hot gas refrigerant. Again, the amount of hot gas flowing through the unit is controlled via the same electronic expansion valve. As with the liquid refrigerant, the hot gas undergoes a phase change (condensation), which releases heat energy into the space. These are called <u>heat pump</u> system. Heat pumps provide both heating and cooling from the same unit and due to added heat of compression, the efficiency of heat pump in heating mode is **higher** compared to the cooling cycle.

Expansion valve is the component that controls the rate at which liquid refrigerant can flow into an evaporator coil. The conventional refrigeration cycle uses "thermostatic expansion valve (TXV)" that uses mechanical spring for control. It has its drawbacks.

- TXV operation is totally independent of compressor operation;
- TXV is susceptible to hunting i.e. overfeeding and starvation of refrigerant flow to the evaporator.

As evaporator load increases, available refrigerant will boil off more rapidly. If it is completely evaporated prior to exiting the evaporator, the vapor will continue to absorb heat (superheat). Although superheating ensures total evaporation of the liquid refrigerant before it goes into the compressor, the density of vapor which quits the evaporator and enters the compressor is reduced leading to reduced refrigeration capacity. The inadequate or high super heat in a system is a concern.

- Too little: liquid refrigerant entering compressor washes out the oil causing premature failure.
- Too much: valuable evaporator space is wasted and possibly causing compressor overheating problems.

The shortcomings of TXV are offset by modern electronic expansion valve.

HVAC - Variable Refrigerant Flow (VRF) Systems

Electronic Expansion Valve (EEV)

With an electronic expansion valve (EEV), you can tell the system what superheat you want and it will set it up. The primary characteristic of EEV is its ability to rotate a prescribed small angle (step) in response to each control pulse applied to its windings. EEV consists of a synchronous electronic motor that can divide a full rotation into a large number of steps, 500 steps/rev. With such a wide range, EEV valve can go from full open to totally closed and closes down when system is satisfied.

EEV in VRF system functions to maintain the pressure differential and also distribute the precise amount of refrigerant to each indoor unit. It allows for the fine control of the refrigerant to the evaporators and can reduce or stop the flow of refrigerant to the individual evaporator unit while meeting the targeted superheat.

Design Considerations for VRF Systems

Deciding what HVAC system best suits your application will depend on several variables viz. building characteristics; cooling and heating load requirements; peak occurrence; simultaneous heating and cooling requirements; fresh air needs; accessibility requirements; minimum and maximum outdoor temperatures; sustainability; and acoustic characteristics.

<u>Building Characteristics</u>

VRF systems are typically distributed systems – the outdoor unit is kept at a far off location like the top of the building or remotely at grade level and all the evaporator units are installed at various locations inside the building. Typically the refrigerant pipe-work (liquid and suction lines) is very long, running in several hundred of feet in length for large multi-storied buildings. Obviously, the long pipe lengths will introduce pressure losses in the suction

HVAC - Variable Refrigerant Flow (VRF) Systems

line and unless the correct diameter of pipe is selected, the indoor units will be starved of refrigerant and it will result in insufficient cooling to the end user. So it is very important to make sure that the pipe sizing is done properly – both for the main header pipe as well as the feeder pipes that feed each indoor unit.

The maximum allowable length varies among different manufacturers; however the general guidelines are as follows:

- The maximum allowable vertical distance between an outdoor unit and its farthest indoor unit is 164 ft;
- The maximum permissible vertical distance between two individual indoor units is 49 feet,
- The maximum overall refrigerant piping lengths between outdoor and farthest indoor unit is up to 541 ft.

Note: The longer the lengths of refrigerant pipes, the more expensive the initial and operating costs.

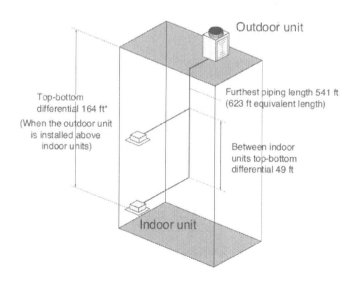

HVAC - Variable Refrigerant Flow (VRF) Systems

Figure Source: ASHRAE

As stated the refrigerant piping criteria varies from manufacturer to manufacture; for example for one of the Japanese manufacturer (Fujitsu), the system design limits are:

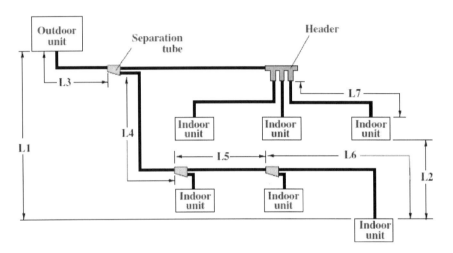

Source: Fujitsu

- L1: Maximum height difference between outdoor unit and indoor unit = 50m

- L2: Maximum height difference between indoor unit and indoor unit = 15m

- L3: Maximum piping length from outdoor unit to first separation tube = 70m

- [L3+L4+L5+L6]: Maximum piping length from outdoor unit to last indoor unit = 100m

- L6 & L7: Maximum piping length from header to indoor unit = 40m

- Total piping length = 200m (Liquid pipe length)

HVAC - Variable Refrigerant Flow (VRF) Systems

Building Load Profile

When selecting a VRF system for a new or retrofit application, the following assessment tasks should be carried out:

- Determine the functional and operational requirements by assessing the cooling load and load profiles including location, hours of operations, number/type of occupants equipment being used etc.

- Determine the required system configuration in terms of the number of indoor units and the outdoor condensing unit capacity by taking into account the total capacity and operational requirement, reliability and maintenance considerations

Building load profile helps to determine the outdoor condensing unit compressor capacity. For instance, if there are many hours at low load, it is advantageous to install multiple compressors and with at least one with inverter (speed adjustment) feature. Figure below shows a typical load profile for an office building.

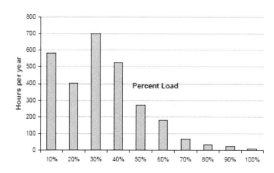

Typical Load Profile of an Office Building

The combined cooling capacity of the indoor sections can match, exceed, or be lower than the capacity of the outdoor section connected to them. But as a normal practice:

- The indoor units are typically sized and selected based on the **greater** of the heating or cooling loads in a zone it serves i.e. maximum peak load expected in any time of the year.

- The outdoor condensing unit is selected based on the load profile of the facility which is the peak load of all the zones combined at any one given time. The important thing here is that it is unlikely all zones will peak at a given time so an element of diversity is considered for economic sizing. Adding up the peak load for each indoor unit and using that total number to size the outdoor unit will result in an unnecessarily oversized condensing unit. Although an oversized condensing unit with multiple compressors is capable of operating at lower capacity, too much over sizing sometimes reduces or ceases the modulation function of the expansion valve. As a rule of thumb, an engineer can specify an outdoor unit with a capacity anywhere between 70% and 130% of the combined capacities of indoor units.

Example:

Say for example a commercial office building with 3 zones. Zone #1 has a peak load of 3 tons, Zone #2 has a peak load of 6 tons and Zone #3 has a peak load of 7 tons.

Combined zone load = 3 + 6 + 7 = 16 tons

Building Peak load = 12 tons

Nearest available sizes for outdoor units = 12.5 ton and 15 ton

Selection outdoor condensing unit of 12.5 tons

Sustainability

HVAC - Variable Refrigerant Flow (VRF) Systems

One attractive feature of the VRF system is its higher efficiency compared to conventional units. Cooling power in VRF system is regulated by means of adjusting the rotation speed of the compressor which can generate an energy saving to the tune of 30%.

VRF system permits easy future expansion as and when the conditions demand. Over sizing however should be avoided unless future expansion is planned.

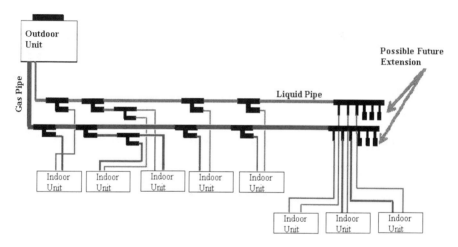

Other sustainability factors include:

- Use of non-ozone depleting environment-friendly refrigerants such as R-410a.

- Opting for heat pump instead of electrical resistance heating in areas demanding both cooling and heating. Heat pumps offer higher energy efficiency.

<u>Simultaneous Heating and Cooling</u>

Some manufacturers offer a VRF system with heat recovery feature, which is capable of providing simultaneous heating and cooling. The cost of VRF-HR is higher than that of normal VRF heat pump units and therefore their application should be carefully evaluated.

More economical design can sometimes be achieved by combining zones with similar heating or cooling requirements together. For examples the areas that may require simultaneous heating and cooling are the perimetric and interior zones. Perimetric areas with lot of glazing and exposure especially towards west and south will have high load variations. VRF heat pump type system is capable of providing simultaneous heating and cooling exceeding 6 tons cooling requirement.

Using VRF heat pump units for heating and cooling can increase building energy efficiency. The designer must evaluate the heat output for the units at the outdoor design temperature. Supplemental heating with electric resistors shall be considered only when the heating capacity of the VRF units is below the heating capacity required by the application. Even though supplement heating is considered, the sequence of operation and commissioning must specify and prevent premature activation of supplemental heating.

First Costs

The installed cost for VRF systems is highly variable, project dependent, and difficult to pin down. Studies indicate that the total installed costs for VRF systems are estimated to be 5% to 20% higher than air or water cooled chilled water system, water source heat pump, or rooftop DX system providing equivalent capacity. This is mainly due to long refrigerant piping and multiple indoor evaporator exchanges with associated controls. Building owners often have no incentive to accept higher first costs even if the claimed payback period is short as the energy savings claims are highly unpredictable.

VRF ADVANTAGES

Comfort

- The main advantage of a variable refrigerant flow (VRF) system is its ability to respond individually to fluctuations in space load conditions. The user can set the ambient temperature of each room as per his/her requirements and the system will automatically adjust the refrigerant flow to suit the requirement;

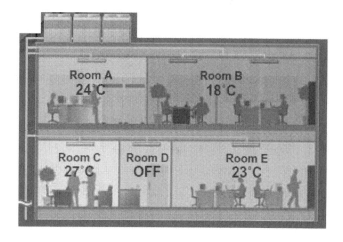

- VRF systems enable wide capacity modulation and bring rooms to the desired temperature extremely quickly and keep temperature fluctuations to minimum. The technology offers excellent dehumidification performance for optimal room humidity regardless of outside conditions. Any area in the building will always be exactly at the right temperature and humidity, ensuring total comfort for their occupants;

HVAC - Variable Refrigerant Flow (VRF) Systems

- VRF systems are capable of simultaneous cooling and heating. Each individual indoor unit can be controlled by a programmable thermostat; most VRF manufacturers offer a centralized control option, which enables the user to monitor and control the entire system from a single location or via the Internet;

- VRF systems can generate separate billing that makes individualized billing easier;

- VRF systems use variable speed compressors (inverter technology) with 10 to 100% capacity range that provides unmatched flexibility for zoning to save energy. Use of inverter technology can maintain precise temperature control, generally within ±1°F.

Design Flexibility

- A single condensing unit can be connected to wide range of indoor units of varying capacity (e.g., 0.5 to 4 tons ducted or ductless configurations such as ceiling recessed, wall-mounted, floor console). Current products enable up to 48 indoor units to be supplied by a single condensing unit;

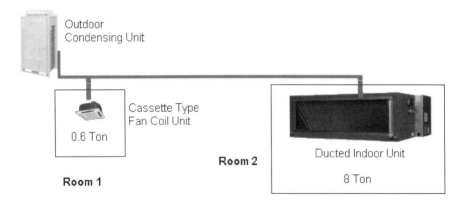

- Modular concept makes it easy to adapt the HVAC system to expansion or reconfiguration of the space. A VRF system can start functioning even while the building/house is still being constructed; unlike large duct or chiller systems, which cannot function until the building project is completed;
- Suitable for putting an outdoor unit on each floor or in mechanical rooms

Flexible Installation

- Enable floor by floor installation and commissioning;
- VRF systems are lightweight; require less outdoor plant space and offers space-saving features;
- Less disruptive to fit in existing buildings (particularly when occupied) and its modular format lends itself to phased installations, and a range of system controller solutions allows the end-user to select the controller that best suits them;
- Modular and self contained units. Multiple of these units can be installed to achieve cooling capacities of hundreds of tons;
- Because ductwork is required only for the ventilation system, it can be smaller than the ducting in standard ducted systems, reducing building height and costs;
- When compared to the single split system, VRF system reduces installation cost by about 30%. VRF system provides reduction in copper tubing and wiring costs.

HVAC - Variable Refrigerant Flow (VRF) Systems

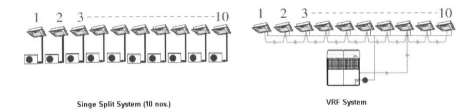

Singe Split System (10 nos.)　　　　VRF System

Energy Efficiency

- VRF systems benefits from the advantages of linear step control in conjunction with inverter and constant speed compressor combination, which allows more precise control of the necessary refrigerant circulation amount required according to the system load. The inverter technology reacts to indoor and outdoor temperature fluctuations by varying power consumption and adjusting compressor speed to its optimal energy usage. Inverter provides superior energy efficiency performance and also allows for a comfortable environment by use of smooth capacity control. Field testing has indicated that this technology can reduce the energy consumption by as much as 30 to 40% a year compared to traditional rotary or reciprocating type compressors.

- VRF technology yields exceptional part-load efficiency. Since most HVAC systems spend most of their operating hours between 30-70% of their maximum capacity, where the coefficient of performance (COP) of the VRF is very high, the seasonal energy efficiency of these systems is excellent.

- A VRF system minimizes or eliminates ductwork completely. This reduces the duct losses often estimated to be 10% to 20% of total airflow in a ducted system.

HVAC - Variable Refrigerant Flow (VRF) Systems

- Inverter compressor technology is highly responsive and efficient. The modular arrangement permits staged operation i.e. Indoor units can easily be turned off in locations needing no cooling, while the system retains highly efficient operation.

- It is possible to include cooling and heating in a single system which avoids duplicating systems (a reversible heat pump only costs 10% more than a cooling unit).

- Energy sub-metering with VRF systems is relatively simple and inexpensive by placing an electric meter on one or a few condensing units. This is a very important feature in the multitenant buildings if energy costs are charged explicitly to each tenant rather than being hidden in overall leasing costs.

Reduced Noise Levels

Indoor and outdoor units are so quiet that they can be placed just about anywhere, giving you more flexibility on how to use indoor and outdoor space. Indoor ductless operating sound levels are as low as 27dB(A) and ducted units sound levels are as low as 29dB(A)

Outdoor units can even be placed directly under a window and quiet indoor units are perfect in environments that require minimal disruption like schools, places or worship, libraries and more. When compared to the single split system, VRF system reduces outside noise levels by almost 5 dB@1m.

HVAC - Variable Refrigerant Flow (VRF) Systems

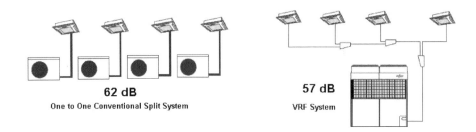

62 dB
One to One Conventional Split System

57 dB
VRF System

Figure Source: Fujitsu

Reliability

Continuous operation is possible even if trouble occurs at an indoor unit.

Indoor Unit

Each indoor unit is controlled individually on the system network. This allows all indoor units continue to run unaffected even if trouble should occur at any indoor unit(s) in one system.

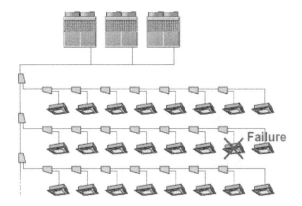

Outdoor Unit

Continuous operation is possible even in the event of compressor failure. There is no immediate system shutdown if trouble occurs in any compressor. The other compressors continue to operate on an emergency basis.

Maintenance and Commissioning

- VRF systems with their standardized configurations and sophisticated electronic controls are aiming toward near plug-and-play commissioning. Normal maintenance for a VRF system is similar to that of any DX system and consists mainly of changing filters and cleaning coils.

- Because there are no water pumps to maintain or air ducts to be cleaned, less maintenance is required compared to other technologies.

Aesthetics

Indoor units are available in different capacities and multiple configurations such as wall-mounted, ceiling-mounted cassette suspended, and concealed ducted types. It is possible to provide an assorted arrangement that combine multiple types of indoor sections with a single outdoor section. These provide extreme versatility to the aesthetic requirements of different building types. Outdoor units can be located on roof or hidden space.

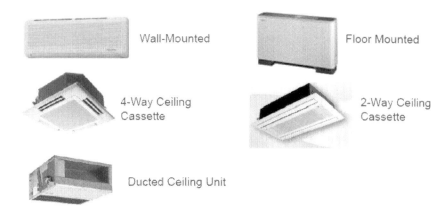

HVAC - Variable Refrigerant Flow (VRF) Systems

Applications

VRF systems may be a particularly good option for buildings with multiple zones or wide variance heating/cooling loads across many different internal zones. These systems provide individual control and are the most versatile of the multi-split systems. Hotels, schools, and office buildings are good examples.

Availability

1) VRF outdoor units can have cooling and heating capacities from 12,000 Btu/h to 300,000 Btu/h;
2) VRF indoor units can have cooling and heating capacities from 5,000 Btu/h to 120,000 Btu/h;
3) The outdoor unit may support up to 48 indoor evaporator units with capacities that collectively add up to 130% capacity of the condensing unit.

VRF equipment is divided into three general categories: residential, light commercial, and applied.

- Residential equipment is single-phase with a cooling capacity of 65,000 Btu/h or less.
- Light commercial equipment is generally three-phase, with cooling capacity greater than 65,000 Btu/h and is designed for small businesses and commercial properties.
- Applied equipment has cooling capacities higher than 135,000 Btu/h and is designed for large commercial buildings.

VRF CHALLENGES & LIMITATIONS

HVAC - Variable Refrigerant Flow (VRF) Systems

VRF systems are not suitable for all applications. The key challenges include:

Refrigerant Piping

VRF system being the split installation is restricted by distance criteria between the condensing unit and the evaporator. The maximum lengths of refrigerant pipework for VRF or any other split systems is determined by the compressors ability to overcome the pressure drop and for the system to maintain proper oil return. All 'split' systems therefore have a maximum vertical and total refrigeration pipework length allowable. This is a considerable disadvantage compared with hydraulic systems which are pumped and as the pump may be sized to suit the system, then theoretically, the hydraulic pipework may be run almost infinite distances. It is important the designer / building owner is aware of these limitations.

Each manufacturer specifies both the size of the pipework required for their system and the maximum permissible vertical and total refrigerant pipework runs. Caution: Although few manufacturers' literature states the refrigerant lines can be as long as 500 feet, but when you read the fine print, after the first 'Tee' from the condensing unit, you are limited to 135 feet to the furthest unit.

Compliance with ANSI/ASHRAE Standard 15-2001

VRF systems must comply with ASHRAE Standard 15-2011—Safety Standard for Refrigeration Systems (ANSI approved). ASHRAE Standard 15-2001 guides designers on how to apply a refrigeration system in a safe manner, and provides information on the type and amount of refrigerant allowed in an *occupied space*.

VRF systems raise the specter of refrigerant leaks, which can be difficult to find and repair, particularly in inaccessible spaces. The refrigerant leak especially if

HVAC - Variable Refrigerant Flow (VRF) Systems

the system serves small rooms can cause oxygen depletion. So you need to limit the system size within reasonable limits based on smallest room area served. For e.g. if the room area is 100 sq-ft, you would need to limit the refrigerant quantity under less than about 30 lbs. Compliance to ASHRAE Standard 15-2011 is sometimes difficult especially where a long length of refrigerant piping is involved. The total refrigerant charge in the refrigerant loop must be within the limits prescribed by ASHRAE Standard 15-2001. This is to ensure the safety of occupants if the entire charge is released.

Few VRF manufacturers have developed products and protocols to address the concerns of refrigerant leakage. Typically, all joints are brazed joints with NO flared fittings. Headers and splitters are specifically designed for the products that do not require flaring or changing wall thicknesses.

Oil Management

As the system has a larger spread, the refrigerant pipes traverse long lengths - hence their pressure testing and protection becomes critical. Long refrigerant piping loops also raise concerns about oil return. Typically, each compressor has its own oil separator, which is optimized for the VRF system. Periodically, the VRF goes into oil retrieval mode, during which time the thermostatic expansion valve opens, and the compressor cycles at high pressure to flush oil out of any locations where it has accumulated.

Fresh Air Requirements (Compliance to ANSI/ASHRAE Standard 62.1)

One of the most challenging aspects of designing VRF systems is the introduction of outside air to comply with ANSI/ASHRAE Standard 62.1, Ventilation for Acceptable Indoor Air Quality, and building codes, which recommends typically 15 to 20 CFM of fresh air per person. Like all split

HVAC - Variable Refrigerant Flow (VRF) Systems

systems, VRF systems do not provide ventilation of their own, so a separate ventilation system is necessary.

Ventilation can be integrated with the VRF system in several ways.

A separate ventilation system and conditioning unit could be installed using conventional technology and the VRF system function would be restricted to the recirculation air. Some VRF units have the ability to handle some outside air and could be used accordingly. Bringing the outside air into the room and then conditioning it with the VRF is not recommended except in dry climates where condensation will not create moisture problems. In humid climates, providing preconditioned outside air to each indoor unit ensures good indoor air quality.

Some manufacturers provide a heat recovery unit which provides heat exchange between incoming outside air and the exhaust air from the air conditioned space, independently of the indoor units. With these systems an equal quantity of outside air and exhaust air is supplied and then exhausted from the air conditioned space. The supply and exhaust air passes over a heat exchanger so heat is recovered from the exhaust air and used to heat or cool the outside air. This solution has the limitation that air is introduced to the space at two different temperatures, i.e. that of the indoor unit and that of the heat recovery unit. If possible it is always ideal to introduce outside air to the indoor unit.

Particulate Matter Removal

ASHRAE Standard 62.1-2004 (item 5.9) specifically discusses particulate matter removal and how VRF indoor units can or cannot uphold the requirements.

Particulate matter filters or air cleaners having a minimum efficiency reporting value (MERV) of not less than 6 when rated in accordance with ANSI/ASHRAE

52.2-1999, shall be provided upstream of all cooling coils or others devices with wetted surfaces trough which the air is supplied to an occupied space. The standard filter with 50% efficiency gravimetric test which is MERV 1 or 2 is NOT acceptable.

Note: High MERV rating filters have a higher cost and high pressure drop. These are often suitable for ducted units and some select ductless unit's viz. Ceiling Mounted Cassette Type – double flow, Ceiling Mounted Cassette Type – multi-flow, Ceiling Mounted Built-in Type, Ceiling Mounted Duct Type, Slim Ceiling Mounted Duct Type and Console – Ceiling Suspended Type.

Higher rating filters aren't available for all type of indoor units viz. Wall Mounted Type, Floor Standing Type/Concealed Floor Standing Type and Ceiling Suspended Cassette Type.

Environmental Concerns

Ozone depletion issues have become a global concern and the issue of a high refrigerant charge associated with long refrigerant lines of VRF system is a strong negative for the system. But with new refrigerant developments, advances in charge management, and controls have transformed the technology to some extent. HFC refrigerants - typically R-410-A and R-407-C are commonly used.

VRF systems are proprietary systems

VRF systems are complete, proprietary systems, from the controls right up to the condensing units, refrigerant controllers, and all the system components other than the refrigerant piping. That means users do not have the flexibility to use "anybody's" building control and automation system to run these systems. You'll need a BacNet or Lonworks black box to connect from your building DDC system to the VRF system, and you can only monitor what it's doing, you can't control it.

Reliability and Maintenance

Although suppliers claim that VRF systems are very reliable, contractors and engineers believe that a VRF system with many compressors (e.g. 20 compressors for 100 tons of cooling) is inherently less reliable than a chiller which has a smaller number of compressors (e.g. 1-4 compressors for 100 tons). However, it is also acknowledged as an advantage since, unlike a chiller, a failure of a single compressor would have limited impact on the system's ability to function.

Performance Guarantee

Currently, no approved ARI standard exists for a performance rating of VRF systems. Consequently, manufacturers need to apply for waivers from the Department of Energy to market their products in the U.S. Although these waivers have been granted, new applications need to be submitted for new product groups.

Concluding.....

VRF provides and alternative, realistic choice to traditional central systems. It captures many of the features of chilled water systems, while incorporating the simplicity of DX systems.

Salient Features:

- Refrigerant flow rate is constantly adjusted by an electronic expansion valve in response to load variations as rooms require more or less cooling. Also, if reversible heat pumps are used, the heating output can be varied to match the varying heat loss in a room;

HVAC - Variable Refrigerant Flow (VRF) Systems

- An expansion valve or control valve can reduce or stop the flow of refrigerant to each indoor unit, thus controlling its output to the room;

- This type of system consists of a number of indoor units (up to 48 and varies per the manufacturer) connected to one or more external condensing units;

- The overall refrigerant flow is varied using either an inverter controlled variable speed compressor, or multiple compressors of varying capacity in response to changes in the cooling or heating requirement within the air conditioned space;

- A control system enables switching between the heating and cooling modes if necessary. In more sophisticated versions, the indoor units may operate in heating or cooling mode independently of others;

- VRF uses inverter or scroll compressors; they are efficient and quiet, these are usually hermetically sealed. Small to medium size units may have 2 compressors;

- Refrigeration pipe work up to 500 feet long is feasible;

- Refrigeration pipe work level differences between indoor and outdoor units up to 150 feet is possible;

- Ozone friendly HFC refrigerants; R-410-A and R-407-C are typically used;

- COP's (Coefficient of Performance) may be as high as 3.8;

- Refrigerant liquid lines tend to be about 3/8" diameter and gas lines about 5/8" to ¾" diameter;

- Central control of a VRV system can be achieved by centralized remote controllers.

HVAC - Variable Refrigerant Flow (VRF) Systems

VRV/VRF technology is based on the simple vapor compression cycle but the system capabilities and limitations must be fully understood and evaluated carefully to determine its suitability. Before working with VRV/VRF systems it is strongly recommended that manufacturer's product training be undertaken.

HVAC - Variable Refrigerant Flow (VRF) Systems

QUIZ

1. VRF systems are characterized by their ability:

a. to control the amount of air flow in each room

b. to control the amount of refrigerant flowing to a series of evaporators in a common system

c. to run multiple ducts for simultaneous heating and cooling

d. All of above

2. Which of the following statement is correct?

a. VRF systems allows use of many evaporators of differing capacities and configurations

b. VRF systems allows for individualized comfort control

c. VRF systems with heat recovery allow for simultaneously heat and cool within the same system

d. All of above

3. VRF systems operate on the _____

a. Chilled water distribution

b. Direct expansion (DX) principle

c. Vapour absorption principle

d. Desiccant drying principle

HVAC - Variable Refrigerant Flow (VRF) Systems

4. In a VRF system layout using the separation tube and header, the separation tube is _____ provided after the header.

    a. Always

    b. Never

    c. Often

    d. Any of above

5. When compared to traditional multi-split systems, the VRF systems _____ the refrigerant path and use _____ copper tubing.

    a. Minimize, less

    b. Maximize, longer

    c. Minimize, longer

    d. Maximize, less

6. A VRF system condensing unit can be connected to wide range of indoor units but these needs to be of same capacity.

    a. True

    b. False

7. Which of the following statement is false?

    a. VRF systems use the same refrigerant cycle as conventional split air conditioning systems

HVAC - Variable Refrigerant Flow (VRF) Systems

b. VRF system gives you the ability to continuously control and adjust the flow of refrigerant to different evaporator units

c. VRF system turns OFF or ON completely in response to one master controller

d. None of above

8. In VRF heat pump systems, when the indoor units are in the heating mode they act as _____.

a. Evaporators

b. Condensers

c. Receivers

d. Accumulators

9. Which of the following system is used where there is a need for some of the spaces to be cooled and some of them to be heated during the same period?

a. Multi-split system

b. VRF heat pump system

c. VRF heat recovery system

d. VRF absorption system

10. A thermal zone is referred to a space or group of spaces within a building with:

HVAC - Variable Refrigerant Flow (VRF) Systems

a. low level partitions

b. return duct carried directly above ceiling

c. similar heating and cooling requirements

d. VRF system installation

11. Which of the following statement is false?

a. VRF systems provide ventilation of there own, so a separate ventilation system is not necessary.

b. VRF system installation is constrained by distance criteria between the condensing unit and the evaporator.

c. VRF systems can operate on reverse cycle heat pump mode

d. All of above

12. Which of the following component of VRF system continually vary the flow of refrigerant into an evaporator coil?

a. Condenser

b. Evaporator

c. Accumulator

d. Electronic expansion valve

13. As a general guideline, the maximum allowable vertical distance between an outdoor unit and its farthest indoor unit is approximately ____.

a. 50 feet

b. 100 feet

c. 150 feet

d. 500 feet

14. Which of the following standard provides information on the type and amount of refrigerant allowed in an occupied space?

a. ASHRAE Standard 15

b. ASHRAE Standard 55

c. ASHRAE Standard 62

d. ASHRAE Standard 90

15. In VRF system, the combined cooling capacity of the indoor units can _____ the capacity of the outdoor section connected to them.

a. Match

b. Exceed

c. Be lower than

d. All of above

HVAC - Variable Refrigerant Flow (VRF) Systems

SEE ANSWERS NEXT PAGE......

HVAC - Variable Refrigerant Flow (VRF) Systems

ANSWERS

1 b	6 b	11 a
2 d	7 c	12 d
3 b	8 b	13 c
4 b	9 c	14 a
5 a	10 c	15 d

Made in the USA
Las Vegas, NV
21 June 2022